THE CAMBRIDGE MISCELLANY

XIII

NATURE AND LIFE

NATURE AND LIFE

By

ALFRED NORTH WHITEHEAD

CAMBRIDGE

AT THE UNIVERSITY PRESS

1934

CAMBRIDGE UNIVERSITY PRESS
Cambridge, New York, Melbourne, Madrid, Cape Town,
Singapore, São Paulo, Delhi, Tokyo, Mexico City

Cambridge University Press
The Edinburgh Building, Cambridge CB2 8RU, UK

Published in the United States of America by
Cambridge University Press, New York

www.cambridge.org
Information on this title: www.cambridge.org/9781107692411

© Cambridge University Press 1934

First published 1934
First paperback edition 2011

A catalogue record for this publication is available from the British Library

ISBN 978-1-107-69241-1 Paperback

I

Philosophy is the product of wonder. The effort after the general characterization of the world around us is the romance of human thought. The correct statement seems so easy, so obvious, and yet it is always eluding us. We inherit the traditional doctrine; we can detect the oversights, the superstitions, the rash generalizations of the past ages. We know so well what we mean and yet we remain so curiously uncertain about the formulation of any detail of our knowledge. This word "detail" lies at the heart of the whole difficulty. You cannot talk vaguely about Nature in general. We must fix upon details within Nature and discuss their essences and their types of interconnection. The world around is complex, composed of details. We have to settle upon the primary types of detail in terms of which we endeavour to express our understanding of Nature. We have to analyse and to abstract, and to understand the natural status of our abstractions. At first sight there are sharp-cut classes within which we can sort the various types of things and characters of things which we find in Nature. Every age manages to find modes of classification which seem fundamental starting points for the researches of the special sciences. Each succeeding age discovers that the primary classifications of its predecessors will not work. In this way a doubt is thrown upon all formulations of laws of Nature which assume these classifications as firm starting points. A problem arises. Philosophy is the search for its solution.

For example, we can conceive Nature as composed of permanent things—namely, bits of matter, moving about in space which otherwise is empty. This way of thinking about Nature has an obvious consonance with common-sense observation. There are chairs, tables, bits of rock, oceans, animal bodies, vegetable bodies, planets, and suns. The enduring self-identity of a house, of a farm, of an animal body, is a presupposition of social intercourse. It is assumed in legal theory. It lies at the base of all literature. A bit of matter is thus conceived as a passive fact, an individual reality which is the same at an instant, or throughout a second, an hour, or a year. Such a material, individual reality supports its various qualifications such as shape, locomotion, colour, or smell, etc. The occurrences of Nature consist in the changes in these qualifications, and more particularly in the changes of motion. The connection between such bits of matter consists purely of spatial relations. Thus, the importance of motion arises from its change of the sole mode of interconnection of material things. Mankind then proceeds to discuss these spatial relations and discovers geometry. The geometrical character of space is conceived as the one way in which Nature imposes determinate relations upon all bits of matter which are the sole occupants of space. In itself, space is conceived as unchanging from eternity to eternity, and as homogeneous from infinity to infinity. Thus, we compose a straightforward characterization of Nature, which is consonant to common sense, and can be verified at each moment of our existence. We sit for hours in the same chair, in the same house, with the same animal body. The dimensions of the room are defined by its spatial relations. There are colours, sounds, scents, partly abiding and partly changing. Also, the major facts of change are defined by locomotion of the animal bodies and of the inorganic furniture. Within this general concept of Nature, there have somehow to be interwoven the further concepts of life and mind.

I have been endeavouring to sketch the general common-sense notion of the universe, which about the beginning of the sixteenth century, say in the year A.D. 1500, was in process of formation among the more progressive thinkers of the European population. It was partly an inheritance from Greek thought and from medieval thought. Partly it was based on the deliverance of direct observation, at any moment verified in the world around us. It was the presupposed support supplying the terms in which the answers to all further questions were found. Among these further questions, the most fundamental and the most obvious are those concerning the laws of locomotion, the meaning of life, the meaning of mentality, and the interrelations of matter, life, and mentality. When we examine the procedures of the great men in the sixteenth and seventeenth centuries, we find them presupposing this general common-sense notion of the universe, and endeavouring to answer all questions in the terms it supplies.

I suggest that there can be no doubt but that this general notion expresses large, all-pervading truths about the world around us. The only question is as to how fundamental these truths may be. In other words, we have to ask what large features of the universe cannot be expressed in these terms. We have also to ask whether we cannot find some other set of notions which will explain the importance of this common-sense notion, and will also explain its relations to those other features ignored by the common-sense notion.

When we survey the subsequent course of scientific thought throughout the seventeenth century up to the present day, two curious facts emerge. In the first place, the development of natural science has gradually discarded every single feature of the original common-sense notion. Nothing whatever remains of it, considered as expressing the primary features in terms of which the universe is to be interpreted. The obvious common-sense notion has been entirely destroyed, so far as concerns its function as the basis for all interpretation. One by one, every item has been dethroned.

There is a second characteristic of subsequent thought which is equally prominent. This common-sense notion still reigns supreme in the workaday life of mankind. It dominates the market place, the playgrounds, the law courts, and in fact the whole sociological intercourse of mankind. It is supreme in literature and is assumed in all the humanistic sciences. Thus, the science of Nature stands opposed to the presuppositions of humanism. Where some conciliation is attempted, it often assumes some sort of mysticism. But in general there is no conciliation.

Indeed, even when we confine attention

to natural science, no special science ever is grounded upon the conciliation of presuppositions belonging to all the various sciences of Nature. Each science confines itself to a fragment of the evidence and weaves its theories in terms of notions suggested by that fragment. Such a procedure is necessary by reason of the limitations of human ability. But its dangers should always be kept in mind. For example, the increasing departmentalization of universities during the last hundred years, however necessary for administrative purposes, tends to trivialize the mentality of the teaching profession. The result of this effective survival of two ways of thought is a patchwork procedure.

Presuppositions from the two points of view are interwoven sporadically. Every special science has to assume results from other sciences. For example, biology presupposes physics. It will usually be the case that these loans really belong to the state of science thirty or forty years earlier. The presuppositions of the physics of my boyhood are to-day powerful influences in the mentality of physiologists. Indeed, we do not need even to bring in the physiologists. The presuppositions of yesterday's physics remain in the minds of physicists, although their explicit doctrines taken in detail deny them.

In order to understand this sporadic interweaving of old and new in modern thought, I will recur to the main principles of the old common-sense doctrine, which even to-day is the common doctrine of ordinary life because in some sense it is true. There are bits of matter, enduring self-identically in space which is otherwise empty. Each bit of matter occupies a definite limited region. Each such particle of matter has its own private qualifications—such as its shape, its motion, its mass, its colour, its scent. Some of these qualifications change, others are persistent. The essential relationship between bits of matter is purely spatial. Space itself is eternally unchanging, always including in itself this capacity for the relationship of bits of matter. Geometry is the science which investigates this spatial capacity for imposing relationship upon matter. Locomotion of matter involves change in spatial relationship. It involves nothing more than that. Matter involves nothing more than spatiality, and the passive support of qualifications. It can be qualified, and it must be qualified. But qualification is a bare fact, which is just itself. This is the grand doctrine of Nature as a self-sufficient, meaningless complex of facts. It is the doctrine of the autonomy of physical science. It is the doctrine which in these lectures I am denying.

The state of modern thought is that every single item in this general doctrine is denied, but that the general conclusions from the doctrine as a whole are tenaciously retained. The result is a complete muddle in scientific thought, in philosophic cosmology, and in epistemology. But any doctrine which does not implicitly presuppose this point of view is assailed as unintelligible.

The first item to be abandoned was the set of qualifications which we distinguish in sense-perception—namely, colour, sound, scent, and analogous qualifications. The transmission theories for light and sound introduced the doctrine of secondary qualities. The colour and the sound were no longer in Nature. They are the mental reactions of the percipient to internal bodily locomotions. Thus, Nature is left with bits of matter, qualified by mass, spatial relations, and the change of such relations.

This loss of the secondary qualities was a severe restriction to Nature, for its value to the percipient was reduced to its function as a mere agent of excitement. Also, the derived mental excitement was not primarily concerned with factors in Nature. The colours and the sounds were secondary factors supplied by the mental reaction. But the curious fact remained that these secondary factors are perceived as related by the spatiality which is the grand substratum of Nature. Hume was, I think, the first philosopher who explicitly pointed out this curious hybrid character of our perceptions, according to the current doctrine of the perception of secondary qualities. Though, of course, this hybrid characteristic was tacitly presupposed by Locke when he conceived colour as a secondary quality of the things in Nature. I believe that any cosmological doctrine which is faithful to the facts has to admit this artificial character of sense-perception. Namely, when we perceive the red rose we are associating our enjoyment of red derived from one source with our enjoyment of a spatial region derived from another source. The conclusion that I draw is that sense-perception, for all its practical importance, is very superficial in its disclosure of the nature of things. This conclusion is supported by the character of delusiveness—that is, of illusion—which persistently clings to sense-perception. For example, our perception of stars which years ago may have vanished, our perceptions of images in mirrors or by refraction, our double vision, our visions under the influence of drugs. My quarrel with modern epistemology concerns its exclusive stress upon sense-perception for the provision of data respecting Nature. Sense-perception does not provide the data in terms of which we interpret it.

This conclusion that pure sense-perception does not provide the data for its own interpretation was the great discovery embodied in Hume's philosophy. This discovery is the reason why Hume's treatise will remain as the irrefutable basis for all subsequent philosophic thought.

Another item in the common-sense doctrine concerns empty space and locomotion. In the first place, the transmission of light and sound shows that space apparently empty is the theatre of activities which we do not directly perceive. This conclusion was explained by the supposition of types of subtle matter, namely, the ether, which we cannot directly perceive. In the second place, this conclusion, and the obvious behaviour of gross ordinary matter, show us that the motions of matter are in some way conditioned by the spatial relations of material bodies to each other. It was here that Newton supplied the great synthesis upon which science was based for more than two centuries. Newton's laws of motion provided a skeleton framework within which more particular laws for the interconnection of bodily motions could be inserted. He also supplied one example of such a particular law in his great law of gravitation, which depended upon mutual distances.

Newton's methodology for physics was an overwhelming success. But the forces which he introduced left Nature still without meaning or value. In the essence of a material body—in its mass, motion, and shape—there was no reason for the law of gravitation. Even if the particular forces could be conceived as the accidents of a cosmic epoch, there was no reason in the Newtonian concepts of mass and motion why material bodies should be connected by any stress between them. Yet the notion of stresses, as essential connections between bodies, was a fundamental factor in the Newtonian concept of Nature. What Newton left for empirical investigation was the determination of the particular stresses now existing. In this determination he made a magnificent beginning by isolating the stresses indicated by his law of gravitation. But he left no hint why, in the nature of things, there should be any stresses at all. The arbitrary motions of the bodies were thus explained by the arbitrary stresses between material bodies, conjoined with their spatiality, their mass, and their initial states of motion. By introducing stresses—in particular the law of gravitation—instead of the welter of detailed transformations of motion, he greatly increased the systematic aspect of Nature. But he left all the factors of the system— more particularly, mass and stress—in the position of detached facts devoid of any reason for their compresence. He thus illustrated a great philosophic truth, that a dead Nature can give no reasons. All ultimate reasons are in terms of aim at value. A dead Nature aims at nothing. It is the essence of life that it exists for its own sake, as the intrinsic reaping of value.

Thus, for Newtonians, Nature yielded no

reasons: it could yield no reasons. Combining Newton and Hume we obtain a barren concept, namely, a field of perception devoid of any data for its own interpretation, and a system of interpretation devoid of any reason for the concurrence of its factors. It is this situation that modern philosophy from Kant onward has in its various ways sought to render intelligible. My own belief is that this situation is a *reductio ad absurdum*, and should not be accepted as the basis for philosophic speculation. Kant was the first philosopher who in this way combined Newton and Hume. He accepted them both, and his three critiques were his endeavour to render intelligible this Hume-Newton situation. But the Hume-Newton situation is the primary presupposition for all modern philosophic thought. Any endeavour to go behind it is, in philosophic discussion, almost angrily rejected as unintelligible.

My aim in these lectures is briefly to point out how both Newton's contribution and Hume's contribution are, each in their way, gravely defective. They are right as far as they go. But they omit those aspects of the universe as experienced, and of our modes of experiencing, which jointly lead to the more penetrating ways of understanding. In the recent situations at Washington, D.C., the Hume-Newton modes of thought can only discern a complex transition of sensa, and an entangled locomotion of molecules, while the deepest intuition of the whole world discerns the President of the United States inaugurating a new chapter in the history of mankind. In such ways the Hume-Newton interpretation omits our intuitive modes of understanding.

I now pass on to the influence of modern science in discrediting the remaining items of the primary common-sense notion with which science in the sixteenth century started its career. But in the present-day reconstruction of physics fragments of the Newtonian concepts are stubbornly retained. The result is to reduce modern physics to a sort of mystic chant over an unintelligible universe. This chant has the exact merits of the old magic ceremonies which flourished in ancient Mesopotamia and later in Europe. One of the earliest fragments of writing which has survived is a report from a Babylonian astrologer to the king, stating the favourable days to turn cattle into the fields, as deduced by his observations of the stars. This mystic relation of observation, theory, and practice is exactly the present position of science in modern life, according to the prevalent scientific philosophy.

The notion of empty space, the mere vehicle of spatial interconnections, has been eliminated from recent science. The whole spatial universe is a field of force—or, in other words, a field of incessant activity.

The mathematical formulae of physics express the mathematical relations realized in this activity.

The unexpected result has been the elimination of bits of matter as the self-identical supports for physical properties. At first, throughout the nineteenth century, the notion of matter was extended. The empty space was conceived as filled with ether. This ether was nothing else than the ordinary matter of the original common-sense notion. It had the properties of a jelly, with its continuity, its cohesion, its flexibility, and its inertia. The ordinary matter of common sense then merely represented certain exceptional entanglements in the ether—that is to say, knots in the ether. These entanglements, which are relatively infrequent throughout space, impose stresses and strains throughout the whole of the jelly-like ether. Also, the agitations of ordinary matter are transmitted through the ether as agitations of the stresses and strains. In this way an immense unification was effected of the various doctrines of light, heat, electricity, and energy, which now coalesced into the one science of the ether. The theory was gradually elaborated throughout the nineteenth century by a brilliant group of physicists and mathematicians, French, German, Dutch, Scandinavian, British, Italian, American. The details of their work, and the relative contributions of various individuals, are not to the point here.

The final result is that the activities of the ether are very different from any of the modes of activity which the common-sense analysis ascribes to ordinary matter. If the doctrine of ether be correct, then our ordinary notions of matter are derived from observations of certain average results which cloak the real nature of the activities of ether. The more recent revolution which has culminated in the physics of the present day has only carried one step farther this trend of nineteenth-century science. Its moral is the extreme superficiality of the broad generalizations which mankind acquires on the basis of sense-perception. The continuous effort to understand the world has carried us far away from all those obvious ideas. Matter has been identified with energy, and energy is sheer activity; the passive substratum composed of self-identical enduring bits of matter has been abandoned, so far as concerns any fundamental description. Obviously this notion expresses an important derivative fact. But it has ceased to be the presupposed basis of theory. The modern point of view is expressed in terms of energy, activity, and the vibratory differentiations of space-time. Any local agitation shakes the whole universe. The distant effects are minute, but they are there. The concept of matter presupposed simple location. Each bit of matter was self-contained, localized in a region with a passive, static network of spatial relations, entwined in a uniform relational system from infinity to infinity and from eternity to eternity. But in the modern concept the group of agitations which we term matter is fused into its environment. There is no possibility of a detached, self-contained local existence. The environment enters into the nature of each thing. Some elements in the nature of a complete set of agitations may remain stable as those agitations are propelled through a changing environment. But such stability is only the case in a general, average way. This average fact is the reason why we find the same chair, the same rock, and the same planet, enduring for days, or for centuries, or for millions of years. In this average fact, the time-factor takes the aspect of endurance, and change is a detail. The fundamental fact, according to the physics of the present day, is that the environment with its peculiarities seeps into

the group-agitation which we term matter,
and the group-agitations extend their character to the environment. In truth, the
notion of the self-contained particle of matter, self-sufficient within its local habitation,
is an abstraction. Now an abstraction is nothing else than the omission of part of the
truth. The abstraction is well founded when
the conclusions drawn from it are not vitiated by the omitted truth.

This general deduction from the modern
doctrine of physics vitiates many conclusions drawn from the applications of physics
to other sciences, such as physiology, or
even such as physics itself. For example,
when geneticists conceive genes as the de-
terminants of heredity. The analogy of the
old concept of matter sometimes leads them
to ignore the influence of the particular animal body in which they are functioning.
They presuppose that a pellet of matter remains in all respects self-identical whatever

be its changes of environment. So far as modern physics is concerned, such characteristics may, or may not, effect changes in the genes, changes which are important in certain respects, though not in others. Thus, no *a priori* argument as to the inheritance of characters can be drawn from the mere doctrine of genes. In fact, recently, physiologists have found that genes are modified in some respects by their environment. The presuppositions of the old common-sense view survive, even when the view itself has been abandoned as a fundamental description.

This survival of fragments of older doctrines is also exemplified in the modern use of the term space-time. The notion of space with its geometry is strictly co-ordinated to the notion of material bodies with simple location in space. A bit of matter is then conceived as self-sufficient with the simple location of the region which it occupies. It is just there, in that region where it is; and it can be described without reference to the goings on in any other region of space. The empty space is the substratum for the passive geometrical relationships between material bodies. These relationships are bare, static facts and carry no consequences which are essentially necessary. For example, Newton's law of gravitation expresses the changes of locomotion which are associated with the spatial relations of material bodies with each other. But this law of gravitation does not result from the Newtonian notion of mass combined with the notion of the occupancy of space, together with the Euclidean geometry. None of these notions either singly or in combination give the slightest warrant for the law of gravitation. Neither Archimedes, nor Galileo, by puzzling over these notions, could have derived any suggestion for the gravitational law. According to the doctrine, space was the substratum for the great all-pervading passive relationship of the natural world. It conditioned all the active relationships, but it did not necessitate them.

The new view is entirely different. The fundamental concepts are activity and process. Nature is divisible and thus extensive. But any division, including some activities and excluding others, also severs the patterns of process which extend beyond all boundaries. The mathematical formulae indicate a logical completeness about such patterns, a completeness which boundaries destroy. For example, half a wave tells only half the story. The notion of self-sufficient isolation is not exemplified in modern physics. There are no essentially self-contained activities within limited regions. These passive geometrical relationships between substrata passively occupying regions have passed out of the picture. Nature is a theatre for the interrelations of activities.

All things change, the activities and their interrelations. To this new concept, the notion of space with its passive, systematic, geometric relationship is entirely inappropriate. The fashionable notion that the new physics has reduced all physical laws to the statement of geometrical relations is quite ridiculous. It has done the opposite. In the place of the Aristotelian notion of the procession of forms, it has substituted the notion of the forms of process. It has thus swept away space and matter, and has substituted the study of the internal relations within a complex state of activity. This complex state is in one sense a unity. There is the whole universe of physical action extending to the remotest star cluster. In another sense, it is divisible into parts. We can trace interrelations within a selected group of activities, and ignore all other activities. By such an abstraction we shall fail to explain those internal activities which are affected by changes in the external system which have been ignored. Also, in any fundamental sense, we shall fail to understand the retained activities. For these activities will depend upon a comparatively unchanging systematic environment.

In all discussions of Nature we must remember the differences of scale, and in particular the differences of time-span. We are apt to take modes of observable functioning of the human body as setting an absolute scale. It is extremely rash to extend conclusions derived from observation far beyond the scale of magnitude to which observation was confined. For example, to exhibit apparent absence of change within a second of time tells nothing as to the change within a thousand years. Also, no apparent change within a thousand years tells anything as to a million years; and no apparent change within a million years tells anything about a million million years. We can extend this progression indefinitely. There is no absolute standard of magnitude. Any term in this progression is large compared to its predecessor and is small compared to its successor.

Again, all special sciences presuppose certain fundamental types of things. Here I am using the word "thing" in its most general sense, which can include activities, colours and other sensa, and values. In this sense, a "thing" is whatever we can talk about. A science is concerned with a limited set of various types of things. There is thus, in the first place, this variety of types. In the second place, there is the determination as to what types are exhibited in any indicated situation. For example, there is the singular proposition—"This is green"; and there is the more general proposition—"All those things are green". This type of enquiry is what the traditional Aristotelian logic takes care of. Undoubtedly such enquiries are essential in the initial stage of any science. But every science strives to get beyond it. Unfortunately, owing to the way in which for over two thousand years philosophic thought has been dominated by its background of Aristotelian logic, all attempts to combine the set of special sciences into a philosophic cosmology, giving some understanding of the universe—all these attempts are vitiated by an unconscious relapse into these Aristotelian forms as the sole mode of expression. The disease of philosophy is its itch to express itself in the forms, "Some S is P", or "All S is P".

Returning to the special sciences, the third step is the endeavour to obtain quantitative decisions. In this stage the typical questions are, "How much P is involved in S?" and "How many S's are P?" In other words, number, quantity, and measurement have been introduced. A simple-minded handling of these quantitative notions can be just as misleading as undue trust in the Aristotelian forms for propositions.

The fourth stage in the development of the science is the introduction of the notion of pattern. Apart from attention to this concept of pattern, our understanding of Nature is crude in the extreme. For example, given an aggregate of carbon atoms and oxygen atoms, and given that the number of oxygen atoms and the number of carbon atoms are known, the properties of the mixture are unknown until the question of pattern is settled. How much free oxygen is there? How much free carbon? How much carbon monoxide? How much carbon dioxide? The answers to some of these questions, with the total quantities of oxygen and of carbon presupposed, will determine the answer to the rest. But, even allowing for this mutual determination, there will be an enormous number of alternative patterns for a mixture of any reasonable amount of carbon and oxygen. And even when the purely chemical pattern is settled, and when the region containing the mixture is given, there are an indefinite number of regional patterns for the distribution of the chemical substances within the containing region. Thus, beyond all questions of quantity, there lie questions of pattern, which are essential for the understanding of Nature. Apart from a presupposed pattern, quantity determines nothing. Indeed, quantity itself is nothing other than analogy of functions within analogous patterns.

Also, this example, involving mere chemical mixture, and chemical combination, and the seclusion of different substances in different subregions of the container, shows us that the notion of pattern involves the concept of different modes of togetherness. This is obviously a fundamental concept which we ought to have thought of as soon as we started with the notion of various types of fundamental things. The danger of all these fundamental notions is that we are apt to assume them unconsciously. When we ask ourselves any question we will usually find that we are assuming certain types of entities involved, that we are assuming certain modes of togetherness of these entities, and that we are even assuming certain widely spread generalities of pattern. Our attention is concerned with details of pattern, and measurement, and proportionate magnitude. Thus, the laws of Nature are merely all-pervading patterns of behaviour, of which the shift and discontinuance lie beyond our ken. Again, the topic of every science is an abstraction from the full concrete happenings of natures. But every abstraction neglects the influx of the factors omitted into the factors retained. Thus, a single pattern discerned by vision limited to the abstractions within a special science differentiates itself into a subordinate factor in an indefinite number of wider patterns when we consider its possibilities of relatedness to the omitted universe. Even within the circle of the special science we may find diversities of functioning not to be explained in terms of that science. But these diversities can be explained when we consider the variety of wider relationships of the pattern in question.

To-day the attitude among many leaders in natural science is a vehement denial of the considerations which have here been put forward. Their attitude seems to me to be a touching example of baseless faith. This judgment is strengthened when we reflect that their position of the autonomy of the natural sciences has its origin in a concept of the world of Nature, now discarded.

Finally, we are left with a fundamental question as yet undiscussed. What are those primary types of things in terms of which the process of the universe is to be understood? Suppose we agree that Nature discloses to the scientific scrutiny merely activities and process. What does this mean? These activities fade into each other. They arise and then pass away. What is being enacted? What is effected? It cannot be that these are merely the formulae of the multiplication table—in the words of a great philosopher, merely a bloodless dance of categories. Nature is full-blooded. Real facts are happening. Physical Nature, as studied in science, is to be looked upon as a complex of the more stable interrelations between the real facts of the real universe.

This lecture has been confined to Nature under an abstraction in which all reference to life was suppressed. The effect of this abstraction has been that dynamics, physics, and chemistry were the sciences which guided our gradual transition from the full common-sense notions of the sixteenth century to the concept of Nature suggested by the speculative physics of the present day. This change of view, occupying four centuries, may be characterized as the transition from space and matter as the fundamental notions to process conceived as a complex of activity with internal relations between its various factors. The older point of view enables us to abstract from change and to conceive of the full reality of Nature *at an instant*, in abstraction from any temporal duration and characterized as to its interrelations solely by the instantaneous distribution of matter in space. According to the Newtonian view, what had thus been omitted was the change of distribution at neighbouring instants. But such change was, on this view, plainly irrelevant to the essential reality of the material universe at the instant considered. Locomotion, and change of relative distribution, was accidental and not essential. Equally accidental was endurance.

Nature at an instant is, in this view, equally real whether or no there be no Nature at any other instant—or, indeed, whether or no there be any other instant. Descartes, who with Galileo and Newton co-operated in the construction of the final Newtonian view, accepted this conclusion. For he explained endurance as perpetual re-creation at each instant. Thus, the matter of fact was, for him, to be seen in the instant and not in the endurance. For him, endurance was a mere succession of instantaneous facts. There were other sides to Descartes' cosmology which might have led him to a greater emphasis on motion. For example, his doctrines of extension and vortices. But in fact, by anticipation, he drew the conclusion which fitted the Newtonian concepts.

There is a fatal contradiction inherent in the Newtonian cosmology. Only one mode of the occupancy of space is allowed for—namely, this bit of matter occupying this region at this durationless instant. This occupation of space is the final real fact, without reference to any other instant, or to any other piece of matter, or to any other region of space. Now, assuming this Newtonian doctrine, we ask—What becomes of velocity at an instant? Again we ask—What becomes of momentum at an instant? These notions are essential for Newtonian physics, and yet they are without any meaning for it. Velocity and momentum require the concept that the state of things at other times and other places enters into the essential character of the material occupancy of space at any selected instant. But the Newtonian concept allows for no such modification of the relation of occupancy. Thus, the cosmological scheme is inherently inconsistent. The mathematical subtleties of the differential calculus afford no help for the removal of this difficulty. We can, indeed, phrase the point at issue in mathematical terms. The

Newtonian notion of occupancy corresponds to the value of a function at a selected point. But the Newtonian physics requires solely the limit of the function at that point. And the Newtonian cosmology gives no hint why the bare fact which is the value should be replaced by the reference to other times and places which is the limit.

For the modern view process, activity, and change are the matter of fact. At an instant there is nothing. Each instant is only a way of grouping matters of fact. Thus, since there are no instants, conceived as simple primary entities, there is no Nature at an instant. Thus, all the interrelations of matters of fact must involve transition in their essence. All realization involves implication in the creative advance.

The discussion in this lecture is only the prolegomenon for the attempt to answer the fundamental question—How do we add content to the notion of bare activity?

Activity for what, producing what, activity involving what?

The next lecture will introduce the concept of life, and will thus enable us to conceive of Nature more concretely, without abstraction.

II

The status of life in Nature is the standing problem of philosophy and of science. Indeed, it is the central meeting point of all the strains of systematic thought, humanistic, naturalistic, philosophic. The very meaning of life is in doubt. When we understand it, we shall also understand its status in the world. But its essence and its status are alike baffling.

After all, this conclusion is not very different from our conclusion respecting Nature, considered in abstraction from the notion of life. We were left with the notion of an activity in which nothing is effected. Also this activity, thus considered, discloses no ground for its own coherence. There is merely a formula for succession. But there is an absence of understandable causation to give a reason for that formula for that succession. Of course, it is always possible to work one's self into a state of complete contentment with an ultimate irrationality. The popular positivistic philosophy adopts this attitude.

The weakness of this positivism is the way in which we all welcome the detached fragments of explanation attained in our present stage of civilization. Suppose that a hundred thousand years ago our ancestors had been wise positivists. They sought for no reasons. What they had observed was sheer matter of fact. It was the development of no necessity. They would have searched for no reasons underlying facts immediately observed. Civilization would never have developed. Our varied powers of detailed observation of the world would have remained dormant. For the peculiarity of a reason is that the intellectual development of its consequences suggests consequences beyond the topics already observed. The extension of observation waits upon some dim apprehension of reasonable connection. For example, the observation of insects on flowers dimly suggests some congruity between the natures of insects and of flowers, and thus leads to a wealth of observation from which whole branches of science have developed. But a consistent positivist should be content with the observed facts—namely, insects visiting flowers. It is a fact of charming simplicity. There is nothing further to be said upon the matter, according to the doctrine of a positivist. At present the scientific world is suffering from a bad attack of muddle-headed positivism, which arbitrarily applies its doctrine and arbitrarily escapes from it. The whole doctrine of life in Nature has suffered from this positivist taint. We are told that there is the routine described in physical and chemical formulae, and that in the process of Nature there is nothing else.

The origin of this persuasion is the dualism which gradually developed in European thought in respect to mind and Nature. At the beginning of the modern period Descartes expresses this dualism with the utmost distinctness. For him, there are material substances with spatial relations, and mental substances. The mental substances are external to the material substances. Neither type requires the other type for the completion of its essence. Their unexplained interrelations are unnecessary for their respective existences. In truth, this formulation of the problem in terms of minds and matter is unfortunate. It omits the lower forms of life, such as vegetation and the lower animal types. These forms touch upon human mentality at their highest, and upon inorganic Nature at their lowest.

The effect of this sharp division between Nature and life has poisoned all subsequent philosophy. Even when the co-ordinate existence of the two types of actualities is abandoned, there is no proper fusion of the two in most modern schools of thought. For some, Nature is mere appearance and mind is the sole reality. For others, physical Nature is the sole reality and mind is an epiphenomenon. Here the phrases "mere appearance" and "epiphenomenon" obviously carry the implication of slight importance for the understanding of the final nature of things.

The doctrine that I am maintaining is that neither physical Nature nor life can be understood unless we fuse them together as essential factors in the composition of "really real" things whose interconnections and individual characters constitute the universe.

The first step in the argument must be to form some concept of what life can mean. Also, we require that the deficiencies in our concept of physical Nature should be supplied by its fusion with life. And we require that, on the other hand, the notion of life should involve the notion of physical Nature.

Now as a first approximation the notion of life implies a certain absoluteness of self-enjoyment. This must mean a certain immediate individuality, which is a complex process of appropriating into a unity of existence the many data presented as relevant by the physical processes of Nature. Life implies the absolute, individual self-enjoyment arising out of this process of appropriation. I have, in my recent writings, used the word "prehension" to express this process of appropriation. Also, I have termed each individual act of immediate self-enjoyment an "occasion of experience". I hold that these unities of existence, these occasions of experience, are the really real things which in their collective unity compose the evolving universe, ever plunging into the creative advance.

But these are forward references to the issue of the argument. As a first approximation we have conceived life as implying absolute, individual self-enjoyment of a process of appropriation. The data appropriated are provided by the antecedent functioning of the universe. Thus, the occasion of experience is absolute in respect to its immediate self-enjoyment. How it deals with its data is to be understood without reference to any other concurrent occasions. Thus, the occasion, in reference to its internal process, requires no contemporary process in order to exist. In fact this mutual independence in the internal process of self-adjustment is the definition of contemporaneousness.

This concept of self-enjoyment does not exhaust that aspect of process here termed "life". Process for its intelligibility involves the notion of a creative activity belonging to the very essence of each occasion. It is the process of eliciting into actual being factors in the universe which antecedently to that process exist only in the mode of unrealized potentialities. The process of self-creation is the transformation of the potential into the actual, and the fact of such transformation includes the immediacy of self-enjoyment.

Thus, in conceiving the function of life in an occasion of experience, we must discriminate the actualized data presented by the antecedent world, the non-actualized potentialities which lie ready to promote their fusion into a new unity of experience, and the immediacy of self-enjoyment which belongs to the creative fusion of those data with those potentialities. This is the doctrine of the creative advance whereby it belongs to the essence of the universe, that it passes into a future. It is nonsense to conceive of Nature as a static fact, even for an instant devoid of duration. There is no Nature apart from transition, and there is no transition apart from temporal duration. This is the reason why the notion of an instant of time, conceived as a primary simple fact, is nonsense.

But even yet we have not exhausted the notion of creation which is essential to the understanding of Nature. We must add yet another character to our description of life. This missing characteristic is "aim". By this term "aim" is meant the exclusion of the boundless wealth of alternative potentiality, and the inclusion of that definite factor of novelty which constitutes the selected way of entertaining those data in that process of unification. The aim is at that complex of feeling which is the enjoyment of those data in that way. "That way of enjoyment" is selected from the boundless wealth of alternatives. It has been aimed at for actualization in that process.

Thus, the characteristics of life are absolute self-enjoyment, creative activity, aim. Here "aim" evidently involves the entertainment of the purely ideal so as to be directive of the creative process. Also, the enjoyment belongs to the process and is not a characteristic of any static result. The aim is at the enjoyment belonging to the process.

The question at once arises as to whether this factor of life in Nature, as thus interpreted, corresponds to anything that we observe in Nature. All philosophy is an endeavour to obtain a self-consistent understanding of things observed. Thus, its development is guided in two ways—one is the demand for a coherent self-consistency, and the other is the elucidation of things observed. It is, therefore, our first task to compare the foregoing doctrine of life in Nature with our direct observations.

Without doubt the sort of observations most prominent in our conscious experience are the sense-perceptions. Sight, hearing, taste, smell, touch constitute a rough list of our major modes of perception through the senses. But there are an indefinite set of obscure bodily feelings which form a background of feeling with items occasionally flashing into prominence. The peculiarity of sense-perception is its dual character, partly irrelevant to the body and partly referent to the body. In the case of sight, the irrelevance to the body is at its maximum. We look at the scenery, at a picture, or at an approaching car on the road, as an external presentation given for our mental entertainment or mental anxiety. There it is, exposed to view. But, on reflection, we elicit the underlying experience that we were seeing with our eyes. Usually this fact is not in explicit consciousness at the moment of perception. The bodily reference is recessive, the visible presentation is dominant. In the other modes of sensation the body is

63

more prominent. There is great variation in this respect between the different modes. In any doctrine as to the information derived from sense-perception this dual reference—external reference and bodily reference—should be kept in mind. The current philosophic doctrines, mostly derived from Hume, are defective by reason of their neglect of bodily reference. Their vice is the deduction of a sharp-cut doctrine from an assumed sharp-cut mode of perception. The truth is that our sense-perceptions are extraordinarily vague and confused modes of experience. Also, there is every evidence that their prominent side of external reference is very superficial in its disclosure of the universe. It is important. For example, pragmatically a paving stone is a hard, solid, static, irremovable fact. This is what sense-perception, on its sharp-cut side, discloses. But if physical science be correct, this is a very superficial account of that portion of the universe which we call the paving stone. Modern physical science is the issue of a co-ordinated effort, sustained for more than three centuries, to understand those activities of Nature by reason of which the transitions of sense-perception occur.

Two conclusions are now abundantly clear. One is that sense-perception omits any discrimination of the fundamental activities within Nature. For example, consider the difference between the paving stone as perceived visually, or by falling upon it, and the molecular activities of the paving stone as described by the physicist. The second conclusion is the failure of science to endow its formulae for activity with any meaning. The divergence of the formulae about Nature from the appearance of Nature has robbed the formulae of any explanatory character. It has even robbed us of reason for believing that the past gives any ground for expectation of the future.

In fact, science conceived as resting on mere sense-perception, with no other source of observation, is bankrupt, so far as concerns its claim to self-sufficiency.

Science can find no individual enjoyment in Nature; science can find no aim in Nature; science can find no creativity in Nature; it finds mere rules of succession. These negations are true of natural science. They are inherent in its methodology. The reason for this blindness of physical science lies in the fact that such science only deals with half the evidence provided by human experience. It divides the seamless coat—or, to change the metaphor into a happier form, it examines the coat, which is superficial, and neglects the body, which is fundamental.

The disastrous separation of body and mind which has been fixed on European thought by Descartes is responsible for this blindness of science. In one sense the abstraction has been a happy one, in that it has allowed the simplest things to be considered first, for about ten generations. Now these simplest things are those widespread habits of Nature that dominate the whole stretch of the universe within our remotest, vaguest observation. None of these laws of Nature gives the slightest evidence of necessity. They are the modes of procedure which within the scale of our observations do in fact prevail. I mean the fact that the extensiveness of the universe is dimensional, the fact that the number of spatial dimensions is three, the spatial laws of geometry, the ultimate formulae for physical occurrences. There is no necessity in any of these ways of behaviour. They exist as average, regulative conditions because the majority of actualities are swaying each other to modes of interconnection exemplifying those laws. New modes of self-expression may be gaining ground. We cannot tell. But, to judge

67 5-2

by all analogy, after a sufficient span of existence our present laws will fade into unimportance. New interests will dominate. In our present sense of the term, our spatio-physical epoch will pass into that background of the past, which conditions all things dimly and without evident effect on the decision of prominent relations.

These massive laws, at present prevailing, are the general physical laws of inorganic Nature. At a certain scale of observation they are prevalent without hint of interference. The formation of suns, the motions of planets, the geologic changes on the earth, seem to proceed with a massive impetus which excludes any hint of modification by other agencies. To this extent sense-perception on which science relies discloses no aim in Nature.

Yet it is untrue to state that the general observation of mankind, in which sense-perception is only one factor, discloses no aim. The exact contrary is the case. All explanations of the sociological functionings of mankind include "aim" as an essential factor in explanation. For example, in a criminal trial where the evidence is circumstantial the demonstration of motive is one chief reliance of the prosecution. In such a trial would the defence plead the doctrine that purpose could not direct the motions of the body, and that to indict the thief for stealing was analogous to indicting the sun for rising? Again no statesman can conduct international relations without some estimate—implicit or explicit in his consciousness—of the types of patriotism respectively prevalent in various nations and in the statesmen of these nations. A lost dog can be seen trying to find his master or trying to find his way home. In fact we are *directly* conscious of our purposes as *directive* of our actions. Apart from such direction no doctrine could in any sense be acted upon. The notions entertained mentally would have no effect upon bodily actions. Thus, what happens would happen in complete indifference to the entertainment of such notions.

Scientific reasoning is completely dominated by the presupposition that mental functionings are not properly part of Nature. Accordingly it disregards all those mental antecedents which mankind habitually presuppose as effective in guiding cosmological functionings. As a method this procedure is entirely justifiable, provided that we recognize the limitations involved. These limitations are both obvious and undefined. The gradual eliciting of their definition is the hope of philosophy.

The points that I would emphasize are: First, that this sharp division between mentality and Nature has no ground in our fundamental observation. We find ourselves living within Nature. Second, I conclude that we should conceive mental operations as among the factors which make up the constitution of Nature. Third, that we should reject the notion of idle wheels in the process of Nature. Every factor which emerges makes a difference, and that difference can only be expressed in terms of the individual character of that factor. Fourth, that we have now the task of defining natural facts, so as to understand how mental occurrences are operative in conditioning the subsequent course of Nature.

A rough division can be made of six types of occurrences in Nature. The first type is human existence, body and mind. The second type includes all sorts of animal life, insects, the vertebrates, and other genera. In fact all the various types of animal life other than human. The third type includes all vegetable life. The fourth type consists of the single living cells. The fifth type consists of all large-scale inorganic aggregates, on a scale comparable to the size of animal bodies, or larger. The sixth type is composed of the happenings on an infinitesimal scale, disclosed by the minute analysis of modern physics.

Now all these functionings of Nature influence each other, require each other, and lead on to each other. The list has purposely been made roughly, without any scientific pretension. The sharp-cut scientific classifications are essential for scientific method, but they are dangerous for philosophy. Such classification hides the truth that the different modes of natural existence shade off into each other. There is the animal life with its central direction of a society of cells, there is the vegetable life with its organized republic of cells, there is the cell life with its organized republic of molecules, there is the large-scale inorganic society of molecules with its passive acceptance of necessities derived from spatial relations, there is the infra-molecular activity which has lost all trace of the passivity of inorganic Nature on a larger scale.

In this survey some main conclusions stand out. One conclusion is the diverse modes of functioning which are produced by diverse modes of organization. The second conclusion is the aspect of continuity between these different modes. There are border-line cases, which bridge the gaps. Often the border-line cases are unstable, and pass quickly. But span of existence is merely relative to our habits of human life. For infra-molecular occurrences, a second is a vast period of time. A third conclusion is the difference in the aspects of Nature according as we change the scale of observation. Each scale of observation presents us with average effects proper to that scale.

Again, another consideration arises. How do we observe Nature? Also, what is the proper analysis of an observation? The conventional answer to this question is that we perceive Nature through our senses. Also, in the analysis of sense-perception we are apt to concentrate upon its most clear-cut instance, namely, sight. Now visual perception is the final product of evolution. It belongs to high-grade animals—to vertebrates and to the more advanced type of insects. There are numberless living things which afford no evidence of possessing sight. Yet they show every sign of taking account of their environment in the way proper to living things. Also, human beings shut off sight with peculiar ease, by closing their eyes or by the calamity of blindness. The information provided by mere sight is peculiarly barren—namely, external regions disclosed as coloured. There is no necessary transition of colours, and no necessary selection of regions, and no necessary mutual adaptation of the display of colours. Sight at any instant merely provides the passive fact of regions variously coloured. If we have memories, we observe the transition of colours. But there is nothing intrinsic to the mere coloured regions which provides any hint of internal activity whereby change can be understood. It is from this experience that our conception of a spatial distribution of passive material substances arises. Nature is thus described as made up of vacuous bits of matter with no internal values, and merely hurrying through space.

But there are two accompaniments of this experience which should make us suspicious of accepting it at its face value as any direct disclosure of the metaphysical nature of things. In the first place, even in visual experience we are also aware of the intervention of the body. We know directly that we see *with our eyes*. That is a vague feeling, but extremely important. Second, every type of crucial experiment proves that what we see, and where we see it, depend entirely upon the physiological functioning of our body. Any method of making our body function internally in a given way will provide us with an assigned visual sensation. The body is supremely indifferent to the happenings of Nature a short way off, where it places its visual sensa.

Now, the same is true of all other modes of sensation, only to a greater extent. All sense-perception is merely one outcome of the dependence of our experience upon bodily functionings. Thus, if we wish to understand the relation of our personal experience to the activities of Nature, the proper procedure is to examine the dependence of our personal experiences upon our personal bodies.

Let us ask about our overwhelming persuasions as to our own personal body-mind relation. In the first place, there is the claim to unity. The human individual is one fact, body and mind. This claim to unity is the fundamental fact, always presupposed, rarely explicitly formulated. I am experiencing and my body is mine. In the second place, the functioning of our body has a much wider influence than the mere production of sense-experience. We find ourselves in a healthy enjoyment of life by reason of the healthy functionings of our internal organs —heart, lungs, bowels, kidneys, etc. The emotional state arises just because they are not providing any sensa directly associated with themselves. Even in sight, we enjoy our vision because there is no eye-strain. Also, we enjoy our general state of life because we have no stomach-ache. I am insisting that the enjoyment of health, good or bad, is a positive feeling only casually associated with particular sensa. For example, you can enjoy the ease with which your eyes are functioning even when you are looking at a bad picture or a vulgar building. This direct feeling of the derivation of emotion from the body is among our fundamental experiences. There are emotions of various types—but every type of emotion is at least modified by derivation from the body. It is for physiologists to analyse in detail the modes of bodily functioning. For philosophy, the one fundamental fact is that the whole complexity of mental experience is either derived or modified by such functioning. Also, our basic feeling is this sense of derivation, which leads to our claim for unity, body and mind.

But our immediate experience also claims derivation from another source, and equally claims a unity founded upon this alternative source of derivation. This second source is our own state of mind directly preceding the immediate present of our conscious experience. A quarter of a second ago we were entertaining such and such ideas, we were enjoying such and such emotions, and we were making such and such observations of external fact. In our present state of mind we are continuing that previous state. The word "continuing" states only half the truth. In one sense it is too weak, and in another sense it overstates. It is too weak because we not only continue, but we claim absolute identity with, our previous state. It was our very identical self in that state of mind, which is, of course, the basis of our present experience a quarter of a second later. In another sense the word "continuing" overstates. For we do not quite continue in our preceding state of experience. New elements have intervened. All of these new elements are provided by our bodily functionings. We fuse these new elements with the basic stuff of experience provided by our state of mind a quarter of a second ago. Also, as we have already agreed, we claim an identification with our body. Thus, our experience in the present discloses its own nature in two sources of derivation, namely, the body and the antecedent experiential functionings. Also, there is a claim for identification with each of these sources. The body is mine, and the antecedent experience is mine. Still more, there is only one ego, to claim the body and to claim the stream of experience. I submit that we have here the fundamental basic persuasion on which we found the whole practice of our existence. While we exist, body and soul are inescapable elements in our being, each with the full reality of our own immediate self. But neither body nor soul possess the sharp observational definition which at first sight we attribute to them. Our knowledge of the body places it as a complex unity of happenings within the larger field of Nature. But its demarcation from the rest of Nature is vague in the extreme. The body consists of the co-ordinated functionings of billions of molecules. It belongs to the structural essence of the body that, in an indefinite number of ways, it is always losing molecules and gaining molecules. When we consider the question with microscopic accuracy, there is no definite boundary to determine where the body begins and external Nature ends. Again, the body can lose whole limbs, and yet we claim identity with the same body. Also, the vital functions of the cells in the amputated limb ebb slowly. Indeed, the limb survives in separation from the body for an immense time compared to the internal vibratory periods of its molecules. Also, apart from such catastrophes, the body requires the environment in order to exist. Thus, there is a unity of the body with the environment, as well as a unity of body and soul into one person.

But in conceiving our personal identity we are apt to emphasize rather the soul than the body. The one individual is that co-ordinated stream of personal experiences which is my thread of life or your thread of life. It is that succession of self-realization, each occasion with its direct memory of its past and with its anticipation of the future. That claim to enduring self-identity is our self-assertion of personal identity.

Yet, when we examine this notion of the soul, it discloses itself as even vaguer than our definition of the body. First, the continuity of the soul—so far as concerns consciousness—has to leap gaps in time. We sleep or we are stunned. And yet it is the same person who recovers consciousness. We trust to memory, and we ground our trust on the continuity of the functionings of Nature, more especially on the continuity of our body. Thus, Nature in general and the body in particular provide the stuff for the personal endurance of the soul. Again, there is a curious variation in the vividness of the successive occasions of the soul's existence. We are living at full stretch with a keen observation of external occurrence; then external attention dies away and we are lost in meditation; the meditation gradually weakens in vivid presentation—we doze; we dream; we sleep with a total lapse of the stream of consciousness. These functionings of the soul are diverse, variable, and discontinuous. The claim to the unity of the soul is analogous to the claim to the unity of the body, and is analogous to the claim to the unity of body and soul, and is analogous to the claim to the community of the body with an external Nature. It is the task of philosophic speculation to conceive the happenings of the universe so as to render understandable the outlook of physical science and to combine this outlook with these direct persuasions representing the basic facts upon which epistemology must build. The weakness of the epistemology of the eighteenth and nineteenth centuries was that it based itself purely upon a narrow formulation of sense-perception. Also, among the various modes of sensation, visual experience was picked out as the typical example. The result was to exclude all the really fundamental factors constituting our experience.

In such an epistemology we are far from the complex data which philosophic speculation has to account for in a system rendering the whole understandable. Consider the types of community of body and soul, of body and Nature, of soul and Nature, or successive occasions of bodily existence, or the soul's existence. These fundamental interconnections have one very remarkable characteristic. Let us ask what is the function of the external world for the stream of experience which constitutes the soul. This world, thus experienced, is the basic fact within those experiences. All the emotions, and purposes, and enjoyments, proper to the individual existence of the soul, are nothing other than the soul's reactions to this experienced world which lies at the base of the soul's existence. Thus, in a sense, the experienced world is one complex factor in the composition of many factors constituting the essence of the soul. We can phrase this shortly by saying that in one sense the world is in the soul.

But there is an antithetical doctrine balancing this primary truth. Namely, our experience of the world involves the exhibition of the soul itself as one of the components within the world. Thus, there is a dual aspect to the relationship of an occasion of experience as one relatum and the experienced world as another relatum. The world is included within the occasion in one sense, and the occasion is included in the world in another sense. For example, I am in the room, and the room is an item in my present experience. But my present experience is what I now am.

But this baffling antithetical relation extends to all the connections which we have been discussing. For example, consider the enduring self-identity of the soul. The soul is nothing else than the succession of my occasions of experience, extending from birth to the present moment. Now, at this instant, I am the complete person embodying all these occasions. They are mine. On the other hand, it is equally true that my immediate occasion of experience, at the present moment, is only one among the stream of occasions which constitutes my soul. Again, the world for me is nothing else than how the functionings of my body present it for my experience. The world is thus wholly to be discerned within those functionings. Knowledge of the world is nothing else than an analysis of the functionings. And yet, on the other hand, the body is merely one society of functionings within the universal society of the world. We have to construe the world in terms of the bodily society, and the bodily society in terms of the general functionings of the world.

Thus, as disclosed in the fundamental essence of our experience, the togetherness of things involves some doctrine of mutual immanence. In some sense or other, this community of the actualities of the world means that each happening is a factor in the nature of every other happening. After all, this is the only way in which we can understand notions habitually employed in daily life. Consider our notion of "causation". How can one event be the cause of another? In the first place, no event can be wholly and solely the cause of another event. The whole antecedent world conspires to produce a new occasion. But some one occasion in an important way conditions the formation of a successor. How can we understand this process of conditioning?

The mere notion of transferring a quality is entirely unintelligible. Suppose that two occurrences may be in fact detached so that one of them is comprehensible without reference to the other. Then all notion of causation between them, or of conditioning, becomes unintelligible. There is—with this supposition—no reason why the possession of any quality by one of them should in any way influence the possession of that quality, or of any other quality, by the other. With such a doctrine the play and interplay of qualitative succession in the world becomes a blank fact from which no conclusions can be drawn as to past, present, or future, beyond the range of direct observation. Such a positivistic belief is quite self-consistent, provided that we do not include in it any hopes for the future or regrets for the past. Science is then without any importance. Also effort is foolish, because it determines nothing. The only intelligible doctrine of causation is founded on the doctrine of immanence. Each occasion presupposes the antecedent world as active in its own nature. This is the reason why events have a determinate status relatively to each other. Also, it is the reason why the qualitative energies of the past are combined into a pattern of qualitative energies in each present occasion. This is the doctrine of causation. It is the reason why it belongs to the essence of each occasion that it is *where* it is. It is the reason for the transference of character from occasion to occasion. It is the reason for the relative stability of laws of Nature, some laws for a wider environment, some laws for a narrower environment. It is the reason why—as we have already noted—in our direct apprehension of the world around us we find that curious habit of claiming a twofold unity with the observed data. We are in the world and the world is in us. Our immediate occasion is in the society of occasions forming the soul, and our soul is in our present occasion. The body is ours, and we are an activity within our body. This fact of observation, vague but imperative, is the foundation of the connexity of the world, and of the transmission of its types of order.

In this survey of the observational data in terms of which our philosophic cosmology must be founded, we have brought together the conclusions of physical science, and those habitual persuasions dominating the sociological functionings of mankind. These persuasions also guide the humanism of literature, of art, and of religion. Mere existence has never entered into the consciousness of man, except as the remote terminus of an abstraction in thought. Descartes' "Cogito, ergo sum" is wrongly translated, "I *think* therefore I am". It is never bare thought or bare existence that we are aware of. I find myself as essentially a unity of emotions, enjoyments, hopes, fears, regrets, valuations of alternatives, decisions—all of them subjective reactions to the environment as active in my nature. My unity—which is Descartes' "I am"—is my process of shaping this welter of material into a consistent pattern of feelings. The individual enjoyment is what I am in my rôle of a natural activity, as I shape the activities of the environment into a new creation, which is myself at this moment; and yet, as being myself, it is a continuation of the antecedent world. If we stress the rôle of the environment, this process is causation. If we stress the rôle of my immediate pattern of active enjoyment, this process is self-creation. If we stress the rôle of the conceptual anticipation of future whose existence is a necessity in the Nature of the present, this process is the teleological aim at some ideal in the future. This aim, however, is not really beyond the present process. For the aim at the future is an enjoyment in the present. It thus effectively conditions the immediate self-creation of the new creature.

We can now again ask the final question as put forward at the close of the former lecture. Physical science has reduced Nature to activity, and has discovered abstract mathematical formulae which are illustrated in these activities of Nature. But the fundamental question remains: How do we add content to the notion of bare activity? This question can only be answered by fusing life with Nature.

In the first place, we must distinguish life from mentality. Mentality involves conceptual experience, and is only one variable ingredient in life. The sort of functioning here termed "conceptual experience" is the entertainment of possibilities for ideal realization in abstraction from any sheer physical realization. The most obvious example of conceptual experience is the entertainment of alternatives. Life lies below this grade of mentality. Life is the enjoyment of emotion, derived from the past and aimed at the future. It is the enjoyment of emotion which was then, which is now, and which will be then. This vector character is of the essence of such entertainment. The emotion transcends the present in two ways. It issues from, and it issues toward. It is received, it is enjoyed, and it is passed along, from moment to moment. Each occasion is an activity of concern, in the Quaker sense of that term. It is the conjunction of transcendence and immanence. The occasion is concerned, in the way of feeling and aim, with things that in their own essence lie beyond it; although these things in their present functions are factors in the concern of that occasion. Thus, each occasion, although engaged in its own immediate self-realization, is concerned with the universe.

The process is always a process of modification by reason of the numberless avenues of supply, and by reason of the numberless modes of qualitative texture. The unity of emotion, which is the unity of the present occasion, is a patterned texture of qualities, always shifting as it is passed into the future. The creative activity aims at preservation of the components and at preservation of intensity. The modifications of pattern, the dismissal into elimination, are in obedience to this aim.

In so far as conceptual mentality does not intervene, the grand patterns pervading the environment are passed on with the inherited modes of adjustment. Here we find the patterns of activity studied by the physicists and chemists. Mentality is merely latent in all these occasions as thus studied. In the case of inorganic Nature any sporadic flashes are inoperative so far as our powers of discernment are concerned. The lowest stages of effective mentality, controlled by the inheritance of physical pattern, involve the faint direction of emphasis by unconscious ideal aim. The various examples of the higher forms of life exhibit the variety of grades of effectiveness of mentality. In the social habits of animals there is evidence of flashes of mentality in the past which have degenerated into physical habits. Finally, in the higher mammals and more particularly in mankind, we have clear evidence of mentality habitually effective. In our own experience, our knowledge consciously entertained and systematized can only mean such mentality, directly observed.

The qualities entertained as objects in conceptual activity are of the nature of catalytic agents, in the sense in which that phrase is used in chemistry. They modify the aesthetic process by which the occasion constitutes itself out of the many streams of feeling received from the past. It is not necessary to assume that conceptions introduce additional sources of measurable energy. They may do so; for the doctrine of the conservation of energy is not based upon exhaustive measurements. But the operation of mentality is primarily to be conceived as a diversion of the flow of energy.

In these lectures I have not entered upon

systematic metaphysical cosmology. The object of the lectures is to indicate those elements in our experience in terms of which such a cosmology should be constructed. The key notion, from which such construction should start, is that the energetic activity considered in physics is the emotional intensity entertained in life.

Philosophy begins in wonder. And, at the end, when philosophic thought has done its best, the wonder remains. There have been added, however, some grasp of the immensity of things, some purification of emotion by understanding. Yet there is a danger in such reflections. An immediate good is apt to be thought of in the degenerate form of a passive enjoyment. Existence is activity ever merging into the future. The aim at philosophic understanding is the aim at piercing the blindness of activity in respect to its transcendent functions.

Milton Keynes UK
Ingram Content Group UK Ltd.
UKHW041520181024
449640UK00009B/76